The Human Body

Translated from the Italian and edited by Maureen Spurgeon

WHAT IS THE HUMAN BODY MADE OF?

The human body is made up of billions of cells, each one able to live and to reproduce on its own. Within the human body there are many types of cells, each type different in shape, structure and size. The 'control centre' of each cell is the nucleus. A jelly-like substance called cytoplasm surrounds the nucleus, and the whole cell is held together by a cellular membrane, which also controls substances going into and out of the cell.

All living things are made up of cells. In turn, a cell is made up of a composition of chemical elements, such as nitrogen, calcium and carbon, but mainly hydrogen and oxygen. These two elements, in the form of water, make up over 70% of the human body.

Cellular
Membrane

• HOW WHY WHEN •

What happens inside the cells?

Each cell absorbs substances from the blood through the cellular membrane. It uses oxygen to 'burn' these substances to produce energy - rather like a car's engine burns petrol in order to make it go.

This energy is used by the cells for the different types of work which they do. Waste substances, such as carbon dioxide, are also produced by the same sort of chemical reaction in the cells.

Each cell can reproduce by dividing itself. In this way, old cells can be replaced.

ORGANELLES

Organelles are the numerous little parts of the cell which float in the cytoplasm. Each organelle has a precise job to do (transformation of substances, production of energy, etc.) The largest organelle is the nucleus.

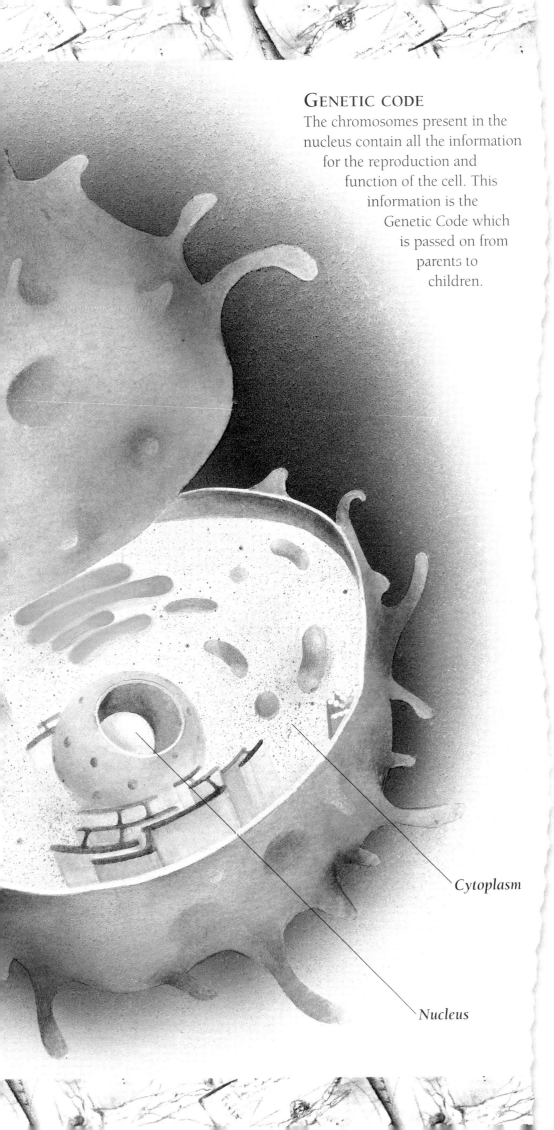

GENETIC CODE

The chromosomes present in the nucleus contain all the information for the reproduction and function of the cell. This information is the Genetic Code which is passed on from parents to children.

Cytoplasm

Nucleus

WHAT IS TISSUE?

A group of cells which have the same characteristics and the same function form a tissue. The epidermal tissue, which forms the top layer of the skin, has a protective function.

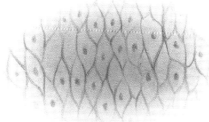

Muscular tissue enables movements.

Connective tissue mainly joins different tissues together.

Nerve tissue enables messages to travel from one part of the body to another, in the form of electrical impulses,

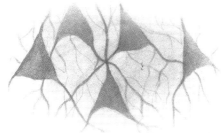

3

WHAT IS THE 'CONTROL CENTRE' OF THE BODY?

Speaking, moving, remembering and reacting to things and substances inside and outside the body, digestion, breathing, feeling hungry, thirsty, pain... all these functions are controlled and regulated by the nervous system, a complicated network of nerve cells called neurons.

The nerves of the peripheral (body) nervous system send information about changes outside the body to the central nervous system, which is the brain and the spinal cord. After receiving and processing information from the peripheral nervous system, the central nervous system responds by sending out the necessary instructions to the body. The central nervous system controls involuntary activity - activity which happens naturally, without us doing anything about it, such as our heartbeat, breathing, contractions of the intestines and the pupils of the eye.

● HOW WHY WHEN ●

Why do we feel hungry and thirsty?

We feel hungry and thirsty when we need to eat or drink because our body has exhausted its reserves. When the stomach is empty and there is a shortage of nutritive substances in the blood, the hypothalamus in the brain causes the sensation of hunger. Similarly, when the level of water in the body is too low, the hypothalamus causes the sensation of thirst. The hypothalamus signals to the pituitary gland, the most important endocrinal (hormone-producing) gland in the body. The pituitary gland then releases a hormone as a message to the kidneys to absorb more water.

WHAT ARE NERVES MADE OF?

Nerves are made of nerve cells or neurons. Each neuron has a cell body of a nucleus and cytoplasm. A series of fibres, like branches, extend from this cell body. The shorter fibres called dendrites pick up nervous impulses which they send to the axon, which sends messages to other nerve cells. An axon can be up to one metre long.

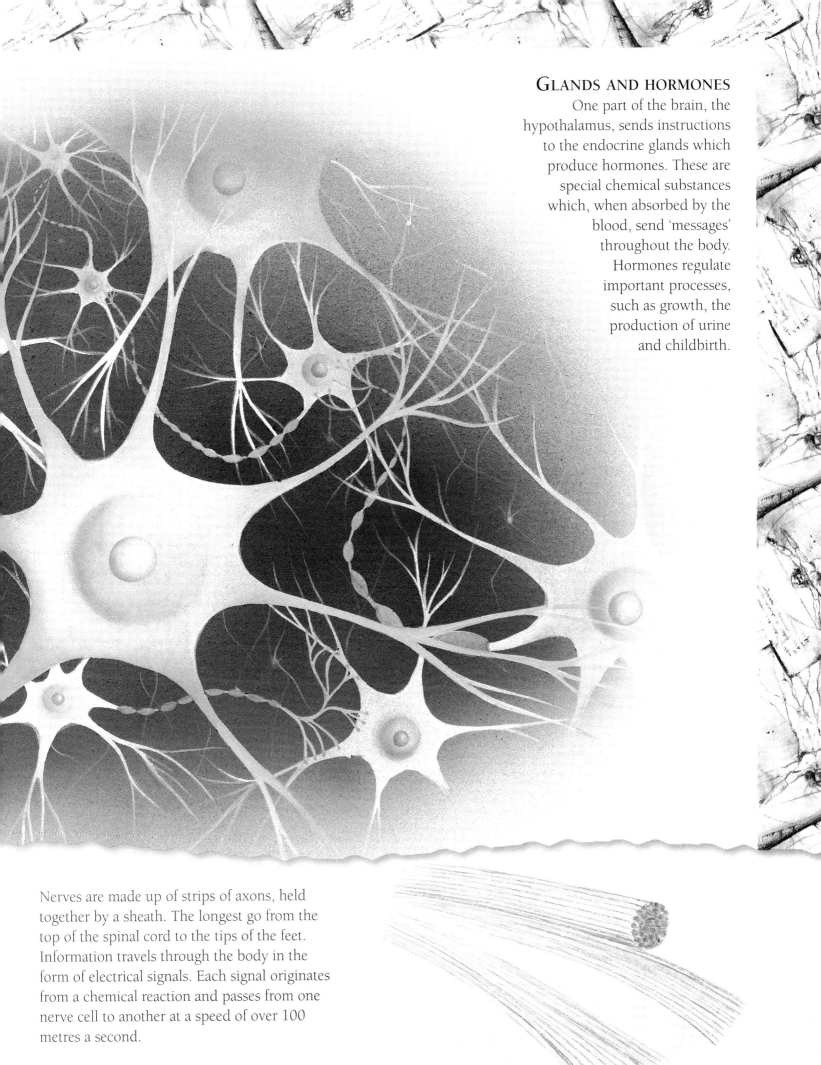

GLANDS AND HORMONES

One part of the brain, the hypothalamus, sends instructions to the endocrine glands which produce hormones. These are special chemical substances which, when absorbed by the blood, send 'messages' throughout the body. Hormones regulate important processes, such as growth, the production of urine and childbirth.

Nerves are made up of strips of axons, held together by a sheath. The longest go from the top of the spinal cord to the tips of the feet. Information travels through the body in the form of electrical signals. Each signal originates from a chemical reaction and passes from one nerve cell to another at a speed of over 100 metres a second.

HOW DOES THE BRAIN WORK?

Complex movements

The brain enables us to be aware of where we are, to communicate, to learn, remember, think and have feelings. It is made up of millions of nerve cells and electrical impulses pass through these cells constantly. A human brain weighs a little less than one kilogram. It gathers, recalls and keeps information gathered by the senses and regulates/controls the function of the body. The brain is divided into two sides or 'hemispheres', right and left, each hemisphere controlling the opposite part of the body and each with specific functions.

The right side co-ordinates movement and controls creative and artistic activity. The left side of the brain controls language, understanding, science and mathematics.

Intelligence and emotions

Language

• HOW WHY WHEN •

Where does the feeling of fear come from?

Fear is an emotion which stems from a tiny area of the brain, the amygdala. When we are faced with menace or danger, this area of the brain causes a chemical reaction which increases not only the heartbeat, but also the rhythm of breathing, the flow of blood to the muscles and perspiration. These reactions are a means of defence, preparing the body to react by fight or by flight. The area of the brain in which emotions begin works closely with the part which controls thinking, so that emotions are kept under control.

THE SPINAL CORD

The spinal cord is a chain of nerve cells, protected by the spinal column. Together with the brain, the spinal cord forms the central nervous system which brings together all activities of the body. Nerves branch off from the spinal cord and these link the central nervous system with the peripheral nervous system.

Taste

Movement Taste

With this simple test you can see how efficient short-term memory is at remembering a limited number of things (say, 5 to 7 things at random).

1) Show this series of letters to a friend for three seconds. Then cover them up and ask your friend to write them in the same order on a piece of paper. Then repeat the test, changing the letters. Who can write the numbers in the correct order each time?

| K | W | V | G | L | R | M |

2) Show ten objects in common use (such as a pencil, a bottle, etc.) to someone for one minute. Then ask the person to list them. Can he or she remember all ten objects?

Visual recognition

HOW DOES THE MEMORY WORK?

Short-term memory enables the retention of information for the time necessary to use it. We use short-term memory to remember a telephone number after having read it.

Long-term memory can gather information permanently. Using long-term memory, we can remember something for a long time, or facts which we remember for studying. By using memory techniques (such as repetition), information can get transferred from short-term to long-term memory.

Sight

THE MEDULLA AND THE PONS

The medulla and the pons are parts of the brain stem. The spinal cord merges into the medulla which is at the bottom of the brain stem. The pons is where nerve fibres come together to connect each side of the body with the opposite hemisphere of the brain.

Balance and muscle co-ordination

DOES THE NOSE SENSE ONLY SMELLS?

The nose does not sense only smells. It also enables us to recognize all the tastes in the food which we eat. When we chew, some smell particles (odour molecules) of food rise up into our nasal cavity, which is connected to the throat. In a small area of this cavity, at eye level, there are numerous little olfactory cells (olfaction means smell). These cells have tiny little hairs called cilia. As soon as odour molecules come in contact with the cilia, the cells send a nerve impulse to the brain, which identifies the taste.

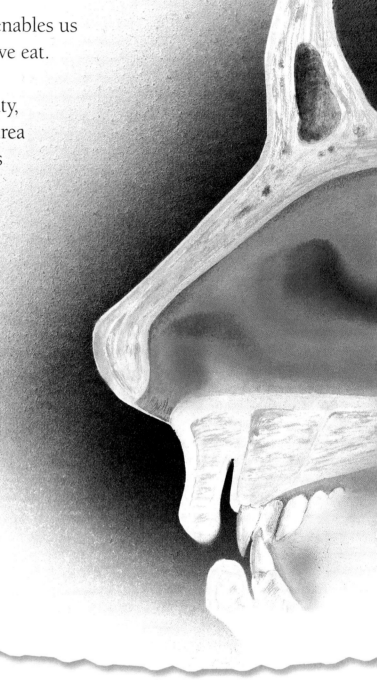

SMELLS IN THE AIR
Smells reach the nose through the air. Substances release molecules which enter the nose. Here the smells are analysed by cells in the nose so that they can then be recognized by the brain.

● HOW WHY WHEN ●

What is the nose made of?
The nose is made up of bone and cartilage. The two openings which allow the passage of air into our body are the nostrils which are lined with fine hairs to filter the air. The nostrils are separated by the nasal septum, a wall of cartilage. Inside, the nose is divided into two nasal passages, which rise up until they form two straight passageways which then connect with the pharynx (the throat). At the top of each nostril are the olfactory cells, which recognize smells and tastes.

HOW MANY TASTES DOES THE TONGUE RECOGNIZE?
The tongue is covered with cells of many different types, called taste receptors, or taste buds. Those at the tip of the tongue distinguish sweet tastes; those along the outer sides salt and sour tastes, and the taste receptors at the bottom of the tongue recognize bitter tastes.

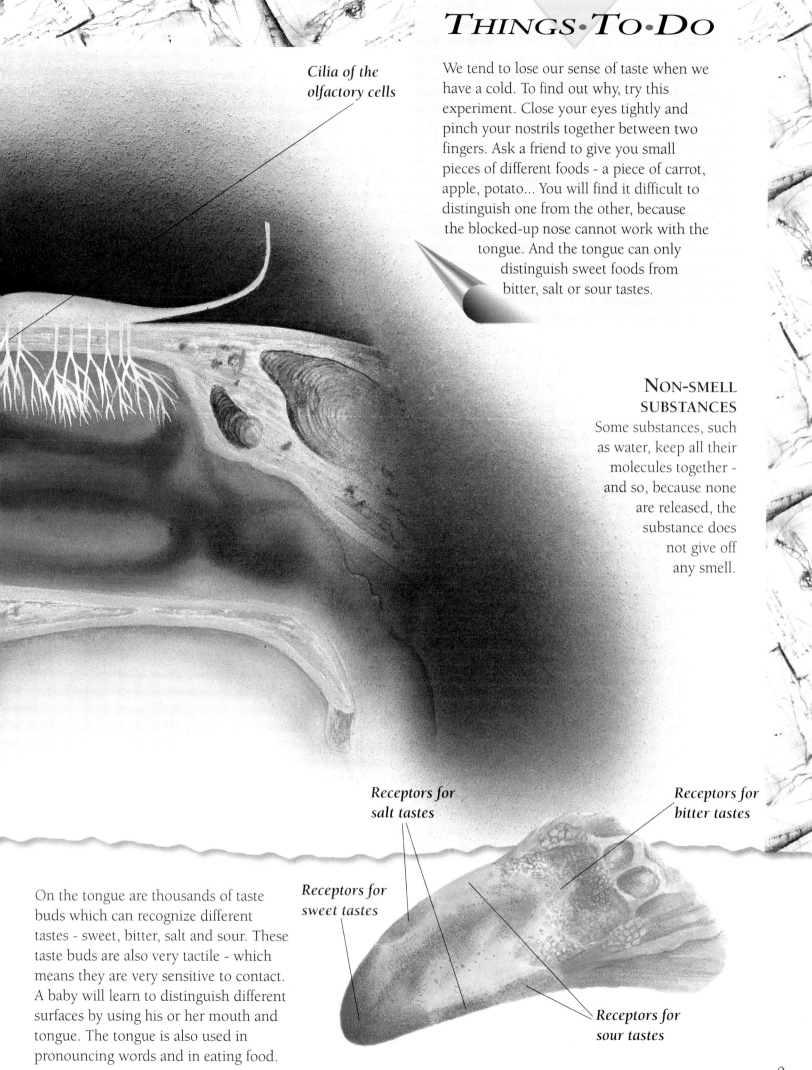

Cilia of the olfactory cells

We tend to lose our sense of taste when we have a cold. To find out why, try this experiment. Close your eyes tightly and pinch your nostrils together between two fingers. Ask a friend to give you small pieces of different foods - a piece of carrot, apple, potato... You will find it difficult to distinguish one from the other, because the blocked-up nose cannot work with the tongue. And the tongue can only distinguish sweet foods from bitter, salt or sour tastes.

NON-SMELL SUBSTANCES

Some substances, such as water, keep all their molecules together - and so, because none are released, the substance does not give off any smell.

Receptors for salt tastes

Receptors for bitter tastes

Receptors for sweet tastes

On the tongue are thousands of taste buds which can recognize different tastes - sweet, bitter, salt and sour. These taste buds are also very tactile - which means they are very sensitive to contact. A baby will learn to distinguish different surfaces by using his or her mouth and tongue. The tongue is also used in pronouncing words and in eating food.

Receptors for sour tastes

WHY CAN'T WE SEE IN THE DARK?

Just as a camera can only capture images on film in the presence of light, so our eyes distinguish shapes and colours of things only when these are hit by a source of light. In the dark, our eyes can only recognize the outline of objects, but not colours. When there is only a little light, our eyes can see a little, because the pupils enlarge (or 'dilate') in order to capture as much light as possible. Scientists have found ways of overcoming the limits of our eyesight by inventing optical instruments, such as binoculars, spectacles and microscopes, and by using infra-red rays. In darkness, infra-red rays can detect the rays of heat given off by a body or by an object and send these back to the eye in the form of luminous rays.

CATS' EYES

Some animals, such as the cat, can see in the dark, because their eyes have lots of tiny cells which can capture even the weakest of light rays and reflect them. That is why we are able to see light shining from the eyes of a cat in the dark.

HOW DOES THE EYE WORK?

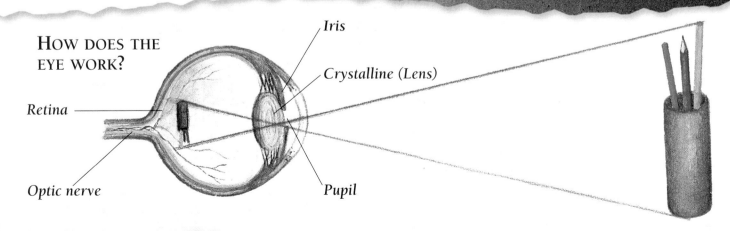

Iris

Crystalline (Lens)

Retina

Optic nerve

Pupil

The brain receives images, it straightens them up, compares, recognizes and memorizes them. In doing all these things, the brain is conditioned by a previous experience, searching for something which it already knows - closing up lines which are apart and imagining shapes which are not there.

Look at these shapes.

How many triangles do you see?

Which letters do you recognize?

In the front part of a human eye we can see the iris (the coloured part) and the pupil (the small black circle at the centre of the eye). When we look at an object, its light is reflected into our eye through the pupil, then through the crystalline or lens by which the eye can focus, before reaching the retina. This acts as a sort of screen, on which the image of the object is received upside down, because the rays of light cross over as they pass through the lens. The cells of the retina transform the light signals into electrical signals and these are sent to the brain through the optic nerve. The brain 'straightens up' the image and so it is recognized.

INDEX

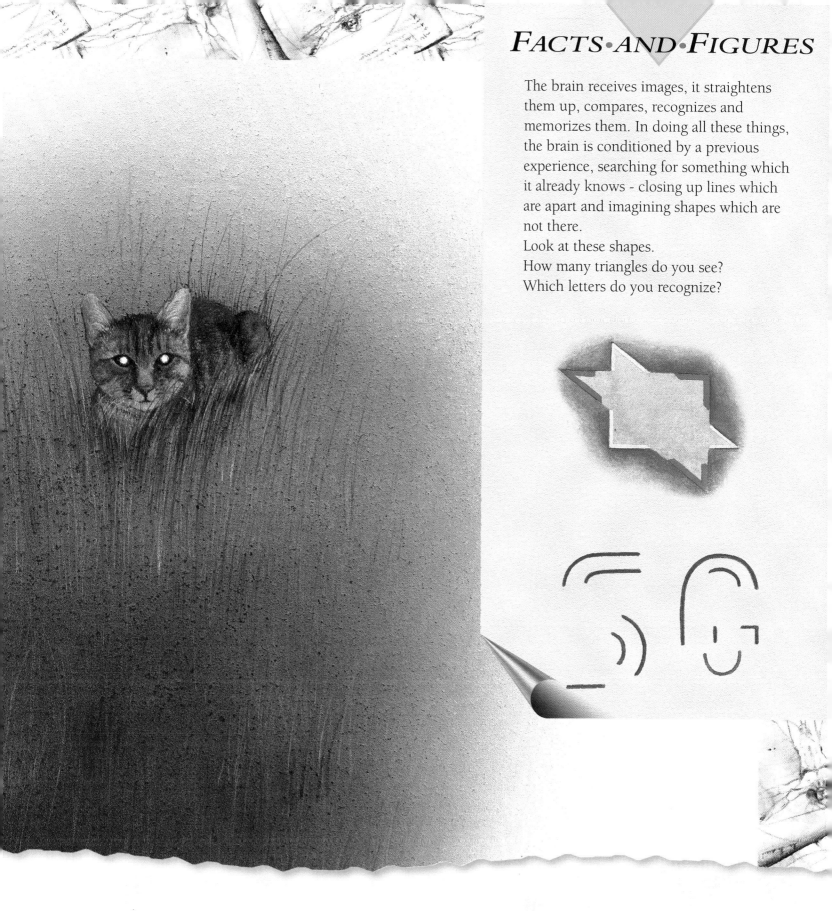

FACTS·AND·FIGURES

The brain receives images, it straightens them up, compares, recognizes and memorizes them. In doing all these things, the brain is conditioned by a previous experience, searching for something which it already knows - closing up lines which are apart and imagining shapes which are not there.

Look at these shapes.
How many triangles do you see?
Which letters do you recognize?

In the front part of a human eye we can see the iris (the coloured part) and the pupil (the small black circle at the centre of the eye). When we look at an object, its light is reflected into our eye through the pupil, then through the crystalline or lens by which the eye can focus, before reaching the retina. This acts as a sort of screen, on which the image of the object is received upside down, because the rays of light cross over as they pass through the lens. The cells of the retina transform the light signals into electrical signals and these are sent to the brain through the optic nerve. The brain 'straightens up' the image and so it is recognized.

WHY DO WE FEEL PAIN?

Skin is important because it covers and protects the human body. It is soft, flexible and waterproof, and made up of many layers. The surface layer is called the epidermis. Beneath the epidermis is the dermis. The dermis has many different sensors which recognize different sensations. The endings of unattached nerves (nerve endings) make us aware of pain, so that if something comes in contact with the skin – an insect bite, or if something hits, scalds or burns us – we can react to it as best we can. The sensitivity of the skin to pain is not the same in all parts of the body. On our fingernails, pain sensitivity is low. Our head is very sensitive to pain. On our arms and legs, there is medium sensitivity.

Epidermis

Sweat gland

● HOW WHY WHEN ●

What use is the skin?

One important job that the skin does is to maintain a constant body temperature of 37°C. When the heat outside the body is excessive, or when the body heats up because of some outer force, the sweat glands in the dermis produce sweat which evaporates and cools the skin. When we are too cold (and also when we are frightened), the skin contracts to try and keep the heat in, and the skin rises up. That is when we see goose-pimples! The deepest layer of skin is made up of a fatty tissue which protects us against cold and from injury.
The skin is soft and stretchy due to sebum, an oily substance produced by the sebaceous glands.

FACTS·AND·FIGURES

In one cubic centimetre of skin there are:

- 65 tiny little muscles connected to as many hairs
- 70 heat receptors
- 15 cold receptors
- 100 sebaceous glands
- more than 500 sweat glands
  - tens of millions of cells
    - The skin is the heaviest organ in the body.
    - The skin of an adult weighs between 4kg - 9kg.

TOUCH

It is the skin which gives us our sense of touch. There are also tactile (touch) receptors inside the body - for example, in the muscles, in bones, in joints and on the tongue.

Sebaceous gland

Vein

Artery

Mass of fatty tissue

SENSITIVE PARTS

The palms of the hands are very sensitive to touch, due to the presence of numerous tactile receptors. The back is less sensitive, because the tactile receptors here are not so close together.

WHAT DO WE FEEL BY TOUCH?

Heat, cold, surfaces which are smooth, rough, light, heavy... each external stimulus releases a sensor in the dermis which sends a message to the brain. Close to the epidermis are the Meissner nerve endings which give us 'fine touch', because they recognize things by the minimum contact with the skin.

Deeper down, the Pacini nerve endings alert us when something presses down on the skin. However, if the pressure of an object is light, such as that of a wrist-watch or of clothes, the nerve endings get used to it and reduce the signals to the brain.

The Ruffini and Krause nerve endings sense an increase and lowering of temperature. These nerve endings make us aware of cold and heat.

WHY ARE OUR SKINS DIFFERENT COLOURS?

Some special cells in the skin contain melanin. This is a pigment (a substance which produces colour) and determines the colour of our skin. Certain types of skin have a lot of melanin and so these types are very dark. Other skin types only have a little melanin, and these are lighter in colour. The quantity of melanin and the consequent colour of the skin is a way of adapting to the surroundings. Dark skins give the best defence against the rays of the Sun. Lighter skins burn more easily.

The palms of the hands and the soles of the feet are the areas of the body which contain the least melanin. Freckles are tiny little areas of the body in which melanin has accumulated.

WHY DOES THE SUN MAKE PEOPLE GO BROWN?

Some rays which contain light from the Sun are ultraviolet (UV) rays. These rays stimulate the production of melanin. When the body is exposed for a length of time in the rays of the Sun, the skin, to avoid these rays damaging it, defends itself by becoming darker. If the exposure to the Sun is too prolonged or if the skin is very fair, the ultraviolet rays burn the skin. Sun cream protects the skin and so avoids the danger of sunburn.

• HOW WHY WHEN •

Is the skin the same all over the body?

The thickness of the skin is not the same all over. For example, the skin is very soft on the eyelids and thicker in the areas which frequently come into contact with outer surfaces - such as the palms of your hands, your fingertips and the soles of your feet.

WHAT IS THE PURPOSE OF OUR SKIN AND OUR HAIR?

The human body is covered all over with hair, except for the palms of the hands and the soles of the feet. There is also hair on some inner parts of the body. Hair gives protection against excessive heat, cold and shock. Each hair begins life in a small sac in the epidermis called a follicle. A hair is alive only in the roots. The part of the hair which we see is made up of dead cells. Curly hair comes from flat follicles, straight hair from round follicles.

Like the skin, the colour of the hair depends on the amount of melanin in the body - there is less melanin in fair-haired people, more in dark-haired.

• Each person loses about 80 hairs every day.

• Uncut fingernails can reach a length of 30cm.

• Each fingernail is completely renewed within six months.

• A person has about 5 million hairs on the body.

• The skin of the eyelid is renewed every six months.

WHY DO WE HAVE FINGERNAILS AND TOE-NAILS?

Nails protect and strengthen the ends of our hands and feet - our fingers and toes. Fingernails also help us to pick up small objects and to scratch! The fingernails and toe-nails are made of a body protein called keratin.
Keratin is also present on the body in the form of hair. It is thick and strong and produced continuously by the skin.
The nails do have a sense of touch. But, like hair, they can be cut without causing us any pain - because we are cutting dead cells. Fingernails grow more quickly than toe-nails.

THE HAIR CYCLE

Hair is constantly renewed in the process of three stages. The first stage is growth in the follicle which lasts two or three years. In the second stage, the roots of the hair rise up towards the outside of the skin. The third stage is when hair is cut or falls out.

WHAT HAPPENS TO THE FOOD WHICH WE EAT?

The food which we ingest (take into) the body will be broken down into smaller and smaller particles, until they can be absorbed by the body cells. The 'journey' begins in the mouth, where the food is chewed by the teeth and moistened by saliva. It is then swallowed into the pharynx (throat) and passes through the oesophagus (gullet) and into the stomach, where it is broken down into more simple substances. Then, as these substances travel through the intestines, they become absorbed through the intestine wall and taken into the blood, which then distributes them around the body. Some substances are used by the cells to reproduce and to nourish them. Other substances give the body the energy necessary for us to move and to think. Those substances which are not needed are passed through the body as waste material - such as faeces and urine.

INSIDE THE INTESTINE
The inside of the small intestine is covered by tiny, little projections called villi. These villi absorb the nutritive substances and pass them out into the blood.

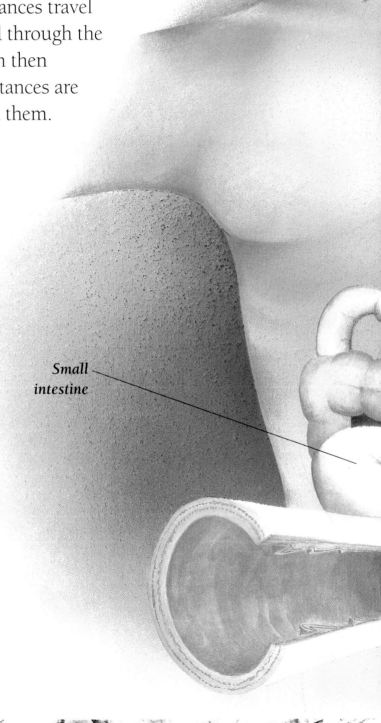

Pharynx

Mouth

Small intestine

● **HOW WHY WHEN** ●

Why do we have different types of teeth?

There are 4 different types of teeth which make up a total of 32 in an adult. The 8 incisors have the job of cutting food up; 4 canine teeth tear the food; the 8 premolars and 12 molars grind it up. Young children only have 20 teeth, which we call 'milk teeth' or 'deciduous teeth' - because when a child is between 5 and 12 years old, these teeth fall out to be replaced by permanent teeth.

FACTS·AND·FIGURES

- Food can remain in the stomach for between 2 and 5 hours, during which time its substances are broken down.
- The small intestine is about 7 metres long.
- There are about 5 million villi in the small intestine.
- On average, one adult eats about 500kg of food each year.

Oesophagus

ro

Stomach

Colon

Rectum

GASTRIC JUICES

During digestion of food, the stomach produces gastric juices. These juices break down the food into more simple substances. Gastric glands in the stomach lining produce a protective mucus, so that the stomach is not damaged by the acid content of the gastric juices.

THE INTESTINE

This is a long, winding tube which is divided into three parts. First is the small intestine, where the process of digestion continues. Second is the large intestine, beginning with the colon which absorbs water and salt from those substances which cannot be digested. The third and last part of the intestine is the rectum, where waste material is passed out of the body in the form of faeces.

WHY DO WE HAVE TO EAT DIFFERENT FOODS?

To keep strong and healthy, the body needs different substances.

Pasta, bread, sugar, fruit and green vegetables provide carbohydrates which give us energy quickly. Oil, butter and fats build up reserves in the tissues which the body transforms into energy more slowly.

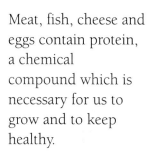

Meat, fish, cheese and eggs contain protein, a chemical compound which is necessary for us to grow and to keep healthy.

Fruit and vegetables provide vitamins to keep us healthy and minerals to renew cells in our bones. These foods also contain fibre, which is needed by the intestines.

Water is essential to the human body. But without food, water is not enough to keep us healthy, however much we drink.

WHAT DOES THE BLOOD DO?

The blood transports and distributes oxygen and nutriments to all the cells in the body. It also keeps the body temperature steady, and combats infection to protect the body from harm. Blood flows through blood vessels, a network of 'canals' thousands of kilometres long, which forms the circulatory (blood circulation) system. The heart pumps the blood around the body through two circuits - through the arteries and through the veins. Through the arteries, the blood, rich in oxygen taken from the lungs and the nutritive substances absorbed through the walls of the intestine, is pumped around to all parts of the body.

Through the veins, the heart pumps the blood to the lungs and back again to the heart. The blood which returns to the heart contains waste substances, such as carbon dioxide, which will be expelled (got rid of) by the body.

Artery

Platelet

Red corpuscles

White corpuscles

• HOW WHY WHEN •

What is the blood made of?

Blood is mostly plasma, a clear liquid. In the plasma are blood corpuscles - red corpuscles, which transport the oxygen and carbon dioxide, and white corpuscles, which defend the body against infection, and the platelets, which makes the blood coagulate (clot) when there is a cut or injury. The liver and the spleen remove old cells from the blood and help to create new blood cells. The blood is also constantly renewed in bone marrow, a tissue inside the bone and connected by blood vessels.

ARTERIES, VEINS AND CAPILLARIES

There are three types of blood vessels - the arteries, thick and strong, the veins, which are thinner and the capillaries, which are thin vessels connecting the veins with the arteries. You can see capillaries on the whites of the eyes.

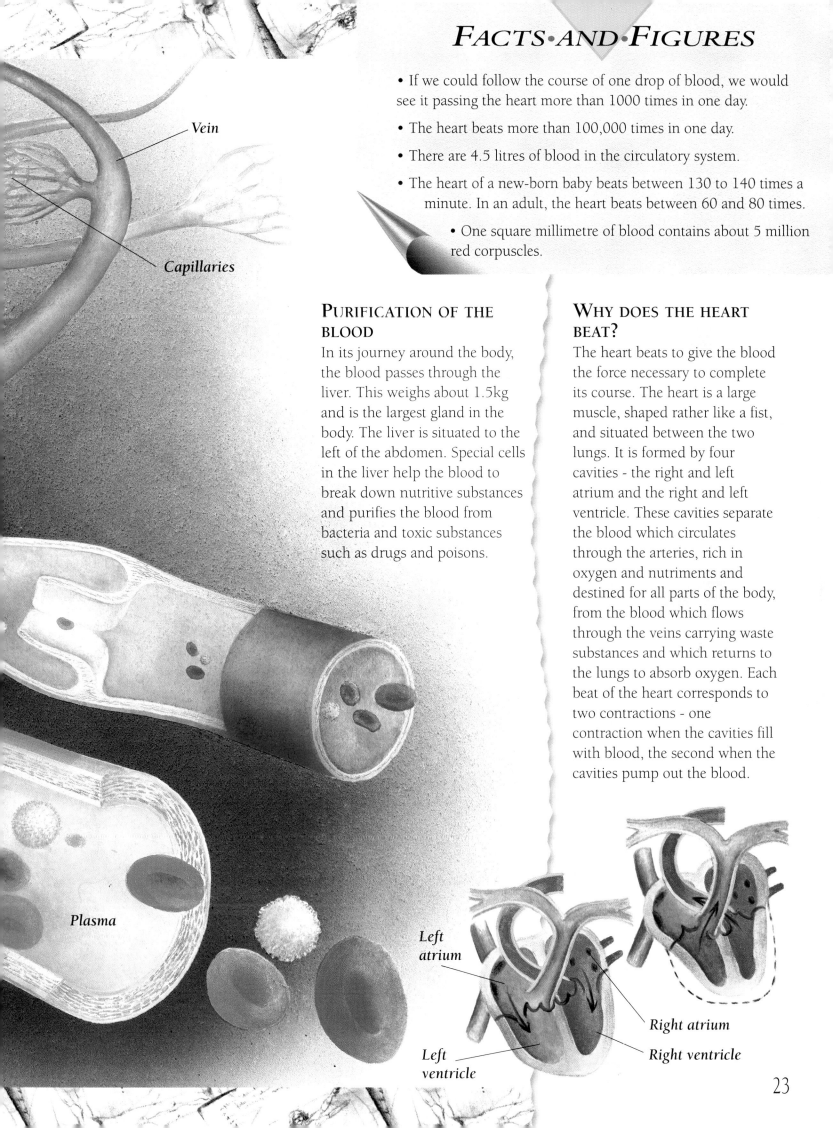

Vein

Capillaries

- If we could follow the course of one drop of blood, we would see it passing the heart more than 1000 times in one day.

- The heart beats more than 100,000 times in one day.

- There are 4.5 litres of blood in the circulatory system.

- The heart of a new-born baby beats between 130 to 140 times a minute. In an adult, the heart beats between 60 and 80 times.

- One square millimetre of blood contains about 5 million red corpuscles.

PURIFICATION OF THE BLOOD

In its journey around the body, the blood passes through the liver. This weighs about 1.5kg and is the largest gland in the body. The liver is situated to the left of the abdomen. Special cells in the liver help the blood to break down nutritive substances and purifies the blood from bacteria and toxic substances such as drugs and poisons.

WHY DOES THE HEART BEAT?

The heart beats to give the blood the force necessary to complete its course. The heart is a large muscle, shaped rather like a fist, and situated between the two lungs. It is formed by four cavities - the right and left atrium and the right and left ventricle. These cavities separate the blood which circulates through the arteries, rich in oxygen and nutriments and destined for all parts of the body, from the blood which flows through the veins carrying waste substances and which returns to the lungs to absorb oxygen. Each beat of the heart corresponds to two contractions - one contraction when the cavities fill with blood, the second when the cavities pump out the blood.

Plasma

Left atrium

Right atrium

Left ventricle

Right ventricle

WHAT IS THE PURPOSE OF THE HUMAN SKELETON?

The skeleton supports the body and protects important organs, such as the brain, heart, lungs and the spinal cord. The bones which comprise the skeleton are strong, but light - because, inside the bones are lots of tiny, empty spaces. These bones are also 'living' and made of body cells which, even when the body has finished developing, are still being renewed. A bone can be long, short, flat, or an irregular shape. Inside most bones there is the marrow, a jelly-like substance in which blood cells form.

- There are about 200 bones in the skeleton of an adult person.

- A bone is six times stronger than a steel bar of the same weight.

- The smallest bone is the stirrup in the inner ear - (2.6mm - 3.4mm). The largest is the femur in the leg (on average about 50cm long).

- We have 27 bones in each hand and 26 in each foot.

- The main part of a bone continues to grow until we are about 20 years old.

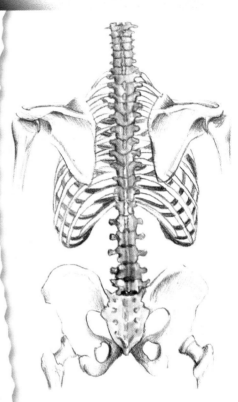

WHY IS THE SPINAL COLUMN FLEXIBLE?
The spinal column is made up of a series of 33 bones called vertebrae and which are connected together. This 'module' structure enables the column to give strong support to the head and the limbs, to keep the body in an erect position. At the same time, the spinal column enables the body to bend, to turn, to stretch and to lean in every direction.

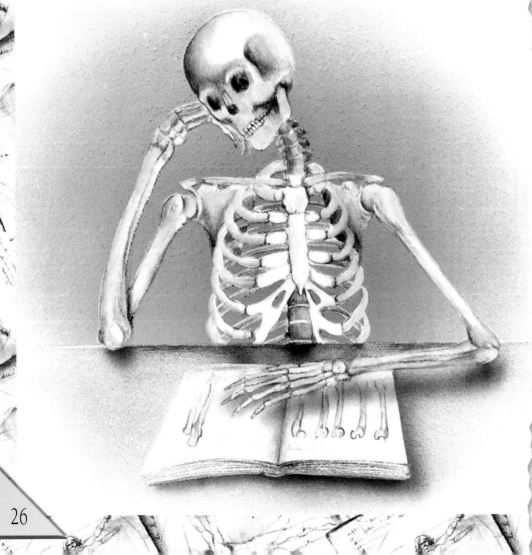

HOW DO OUR BONES ENABLE US TO MOVE?

The bones in our skeleton can move thanks to our joints, which connect the bones, held together by ligaments. The ends of the bones where they rub and press against each other have a smooth protective covering of cartilage, a 'squashy' sort of tissue which makes movement easier. Cartilage also protects the joints and reduces friction. The cartilage ends are protected by an oily liquid called synovial fluid.

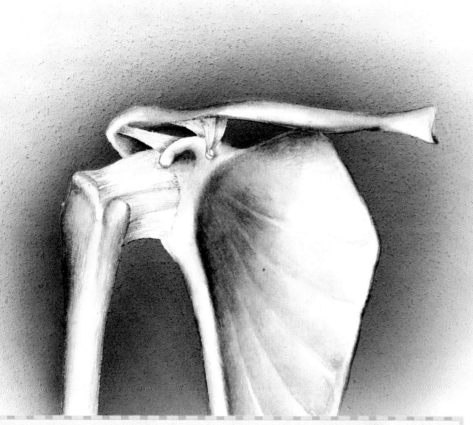

• HOW WHY WHEN •

What are ligaments?

Ligaments are strips of fibre which keep the bone connected to the joint. Some ligaments form strong, complex systems - for example, the ligaments crossing the knees.
Vertebrae are also connected by ligaments which make movements of the spinal column possible.

CARTILAGE STRUCTURE

Some parts of the body, such as the nose and the ears, do not have bone structure. Instead, they have a cartilage structure. Cartilage is a tissue similar to bone, but more flexible and softer. It is present in all joints, as well as the trachea, the bronchi and in vertebral discs.

In a foetus (a developing baby) the skeleton is totally cartilage. It gradually changes into bone tissue. The transformation into bone is completed after birth.

27

WHAT DETERMINES OUR MOVEMENTS?

Playing, jumping, climbing, getting up, sitting down, turning somersaults, laughing or pulling faces... all our movements are made possible by the combined work of the muscles and the bones. The human body has more than 600 muscles of different shapes and sizes. The skeletal muscles are arranged in strips and the cells have a striped pattern. That is why these are called striated (striped) muscles. These are called voluntary muscles, because they work at our command. At the same time, they also control the movements of the bone to which the muscle is connected. Smooth muscles, such as those in the intestines and stomach, work regardless of anything we do. So they are called involuntary muscles. The heart is a special muscle - it is a striated yet involuntary muscle.

INVOLUNTARY REFLEXES

Sometimes the striated muscles also move without a precise 'command' from us. This happens when the hand instinctively withdraws from something which is too hot to handle, or when internal organs jolt from a sense of fear.

• HOW WHY WHEN •

What happens during physical effort?

During intense physical activity, the muscles have to make extra effort. The production of energy needs a larger supply of oxygen and creates lots more carbon dioxide. So the heart has to beat faster to pump more blood, which means that the lungs increase their rhythm and the intensity of respiration in order to circulate the oxygen. Perspiration also increases, in order to get rid of the excess heat generated by the production of energy.

WHY DO SOME MUSCLES WORK TOGETHER?

Some skeletal muscles can 'pull' but not 'push'. So, for some movements, these muscles have to work with others.

TENDONS

Muscles are connected to the bone by ribbon-like fibres called tendons. These can be flat or tube-like and they are very strong. The tendons in the hand can be seen quite easily when we move a finger.

• By the end of a race, the breathing rhythm has increased up to 75 breaths per minute.

• Muscles can represent up to 40% of the total weight of the body.

• There are 40 muscles in the face alone. We use these to make different expressions.

• When we laugh, we use about 20 muscles.

Here we see how the biceps and triceps muscles work in the upper arm. When the biceps muscle contracts, it makes the arm bend and the forearm rises. The triceps is stretched.

When the triceps muscle contracts, it pulls the forearm, which straightens out. Now the biceps muscle is relaxed. Muscles like these which work in a pair are also called antagonistic pairs.

29

INDEX